STORM
CHASERS
THE STUNNING SKIES OF TORNADO ALLEY

RON STENZ

AMERICA
THROUGH TIME®
ADDING COLOR TO AMERICAN HISTORY

America Through Time is an imprint of Fonthill Media LLC
www.through-time.com
office@through-time.com

Published by Arcadia Publishing by arrangement with Fonthill Media LLC
For all general information, please contact Arcadia Publishing:
Telephone: 843-853-2070
Fax: 843-853-0044
E-mail: sales@arcadiapublishing.com
For customer service and orders:
Toll-Free 1-888-313-2665

www.arcadiapublishing.com

First published 2023

Copyright © Ron Stenz 2023

ISBN 978-1-63499-443-9

Typeset in Trade Gothic
Printed and bound in England

CONTENTS

ABOUT THE AUTHOR

RON STENZ is a meteorology professor at the College of DuPage (weather.cod.edu) who leads multiple educationally focused storm chasing trips each spring and summer. His photography focuses on capturing the vivid storms of Tornado Alley for use in the classroom, teaching future meteorologists and storm chasers. As a tornado researcher using supercomputer simulations to try to better understand one of weather's most powerful processes, Ron travels the plains each year in search of violent storms because there is still no substitute for witnessing the true power of nature in person.

INTRODUCTION

This book will take you through the ferocious storms and awe-inspiring skies of "Tornado Alley" through the eyes of a storm chaser and tornado researcher who travels there each year attempting to witness and better understand arguably the most extreme weather on the planet. Ominous skies, imposing thunderstorms, strong tornadoes, destructive winds, damaging hail, and breathtaking landscapes were all part of his storm chasing adventures traveling throughout this unforgiving, yet beautiful region of the world. Infamous for some of the most violent storms on Earth, "Tornado Alley" showcases the power and beauty of supercells and tornadoes traversing the picturesque American landscapes, unleashing their fury on those unfortunate enough to be in the path. Some storms appear imposing and sinister, while others may be so magnificent to view that onlookers find it hard to move away before it is too late to escape the impending danger. The unique geography of the United States allows all the ingredients needed for destructive severe thunderstorms and tornadoes to come together at a frequency greater than that of anywhere else in the world. Despite having no truly agreed upon geographic boundaries, the Great Plains and portions of the Midwest are often included in the colloquialism "Tornado Alley" by storm chasers. See the jaw dropping results as shear, lift, instability, and moisture all come together in abundance above vast and open landscapes to create stunning scenes unlike anywhere else in the world.

A quintessential Tornado Alley scene as a tornado towers over a farm with clouds of dust swirling and rising into the circulation near McCook, Nebraska, during mid-May of 2019. This tornado was ultimately rated an EF-2 after destroying several grain bins and outbuildings, along with significantly damaging a home. Fortunately, there were no fatalities from this strong tornado that was moving across the Nebraska countryside at close to 60 miles per hour with maximum winds estimated near 120 miles per hour.

1

LANDSCAPES

Tornado Alley provides impressive views and memorable scenery even in the absence of powerful thunderstorm activity. Vast stretches of farmland occasionally broken up by small towns are a hallmark of this region, critical to the agricultural industry of the United States. In addition to expansive open lands with the sky seemingly in all directions, more complex terrain such as the Wichita mountains, Nebraska Sandhills, and the Badlands of South Dakota are commonly encountered by storm chasers in pursuit of photogenic storms and tornadoes.

Looking east from atop Mount Scott in the Wichita Mountains of Southwest Oklahoma in May of 2017. This complex terrain in one of the most tornado-prone locations in the country may be a scenic surprise to those unfamiliar with the area.

A large rattlesnake in the Wichita Mountains wildlife refuge in May of 2019. Storm chasers always need to watch their step and be aware of their surroundings, as sights like this are not uncommon throughout Tornado Alley.

A view of Altus Air Force Base in southwest Oklahoma on May 19, 2019. Aircraft were in the process of being evacuated to different bases or stored in safe locations in advance of a rare High Risk convective outlook from the Storm Prediction Center for the next day.

A windmill under turbulent skies in Garfield County, Oklahoma, during May of 2019. Scenes similar to this one can be found throughout the Great Plains.

A pyrocumulus cloud from a wildfire explosively grows in front of the vast, open plains of Colorado on June 23, 2012. Wildfire smoke can end up impacting severe weather setups and visibility up to thousands of miles away.

Above large wheat fields north of Colby, Kansas, crepuscular rays are visible near the edge of a distant thunderstorm during June of 2012.

A colorful sunset above the Mennonite Heritage Park in Henderson, Nebraska, during June of 2021.

Sunset over Lakeview Park in Henderson, Nebraska, during May of 2020. Welcoming and scenic small towns like Henderson are memorable stops even on days without significant thunderstorm activity.

Flattened corn after a tornado in Maple Park, Illinois, in August of 2021. Throughout Tornado Alley, sights like this are occasionally witnessed in the aftermath of tornadoes or strong straight-line winds from storms.

A cumulus tower is growing above the open, hilly grasslands of the Nebraska sandhills in June of 2020. Despite frequent storm activity, this region can be tough for storm chasers due to very limited road options and the rolling, hilly terrain limiting visibility.

Surrounded by seemingly never-ending open grasslands, the South Dakota Badlands are a frequent sightseeing or hiking stop on fair weather days for storm chasers later in the storm season. This view across the dramatic landscape comes on a hot and sunny day in mid-June of 2022.

A view from a helicopter showing the expansive South Dakota Badlands during June of 2015. This unique terrain is the result of deposition and erosion, with the rapid erosion rates for rocks in this area currently estimated at around one inch per year.

Bighorn sheep traverse the slippery terrain of the South Dakota Badlands in May of 2015 shortly after storms moved through.

Cattle blocking dirt and gravel roads is a common sight for storm chasers throughout the Great Plains. In this case, the cattle slowing down the escape from a storm are in Harding County, South Dakota, during mid-June of 2022.

2

HAIL AND WIND

Small particles of ice known as hail embryos can grow into hailstones as large as baseballs, softballs, and even larger when transported high into thunderstorms by an intense updraft. A common misconception is that circulations with large vertical extent are needed to create the large hailstones with a layered appearance often found in Tornado Alley. However, the large size and layering is typically the result of a hailstone passing through regions within the turbulent updraft where hailstones experience "wet growth" and "dry growth" colliding with water droplets on a trajectory that more closely resembles one main trip up and one main trip down inside the storm. "Wet growth" occurs when water takes longer to freeze onto a hailstone, creating transparent layers. "Dry growth" is when water rapidly freezes onto a hailstone trapping air inside the newly formed layer of ice, making it opaque and white in appearance.

Damaging and potentially destructive winds also frequently impact Tornado Alley. Often overlooked compared to tornadoes, "straight-line" winds from thunderstorm downdrafts can occasionally be powerful enough to flip train cars, flatten crops, and even blow roofs off of buildings. These winds are largely driven by the evaporational cooling of air by rain in the downdraft of a storm, which causes this relatively colder and more dense air to accelerate down towards the ground, resulting in gusty winds.

The soft, white area beneath these cumulonimbus clouds indicates where hail is falling from this storm north of Reva, South Dakota, in mid-June of 2022. When a significant tornado threat does not exist, chasers may trail a storm like this for views of its structure and opportunities to collect the falling hail after it has landed on the ground.

A large hailstone south of Buffalo, South Dakota, in mid-June of 2022. Once hail is an inch or larger, about the size of a quarter, it will likely begin to cause damage to vehicles and roofing.

A pile of hailstones from a storm near Buffalo, South Dakota in mid-June of 2022. Comparing hailstones to objects of known size allows quick estimates of hailstone sizes.

Measuring just over 3 inches, this hailstone was found south of Buffalo, South Dakota, in mid-June of 2022. Calipers were used for a more accurate and precise measurement than comparisons to objects of known size.

Accumulated hail melts quickly in heavy rains creating wintry driving conditions near Voss, Texas, in late May of 2022. Even on hot days like this one, a hail core can rapidly create slick and icy roads.

A hail core near Cheyenne Wells, Colorado, in late June of 2019 left a sharp cutoff between green grass seen in the distance along the road and a wintry landscape of hail covered ground.

A thunderstorm tracking along highway 40 east of Cheyenne Wells, Colorado, creates an impressively long stretch of wintry driving conditions in late June of 2019. Hail accumulations and a chilly night led to icy roads, at times even on the asphalt between the swaths of piled up hailstones.

Powerful outflow winds, which surge outwards from a thunderstorm, loft dust across the roadway near Pembina, North Dakota, in August of 2013.

Dirt is lofted thousands of feet high by powerful straight-line winds near Dell Rapids, South Dakota, in early June of 2020. Occasional gustnadoes, which are brief and shallow nontornadic circulations, occurred within this imposing wall of lofted dirt.

Strong winds create an ominous wall of dirt sweeping through a wind farm near Woodstock, Minnesota, in early June of 2020.

Trees are bent and dust races across the road as intense outflow crosses the highway in open country northwest of Terry, Montana, in late June of 2016.

Sehj, a doctor and bioinformaticist, braves strong winds from a storm northwest of Terry, Montana, in late June of 2016 to take photos on his first storm chase. Experiencing the elements is often a thrilling part of the adventure for many storm chasers.

3

STORM STRUCTURE

For some storm chasers, the wondrous cloud formations that frequently develop from the extreme weather over Tornado Alley can be as big of, if not a bigger attraction than witnessing a tornado. Wind shear—the change in wind speed and/or direction with height in the atmosphere—allows the organization of storms and sculpting of clouds into majestic structures that truly need to be seen to be believed. Often, the rain-cooled air of the storms themselves can create additional conditions leading to impressive cloud features, sometimes seemingly stretching the entire horizon.

A large supercell near Burlington, Colorado, in late May of 2017.

An approaching storm near Richardton, North Dakota, in late May of 2018.

Outflow from a storm near Utica, Kansas, in late May of 2017 creates a broad shelf cloud stretching across much of the horizon above pump jacks in a fallow field.

This shelf cloud represents the leading edge of gusty winds near Utica, Kansas, in late May of 2017. Along with the gusty winds come a blast of colder air and dust lifted from the fallow field.

Atop this late June 2015 supercell near Binford, North Dakota, crepuscular rays emerge. Despite the somewhat tranquil appearance, this storm produced both large hail and a tornado.

This mid-morning elevated supercell tracked over 100 miles across northwest Kansas, seen here near Clayton, Kansas, in late May of 2021. An outflow boundary from this storm was likely an important factor for tornadic thunderstorms later in the day over southwest Nebraska and northwest Kansas.

Mammatus clouds bulge from the underside of the anvil of this sculpted supercell near Arnold, Nebraska, in early June of 2020.

Dust kicks up from strong outflow under the shelf cloud of this early June 2020 supercell near Arnold, Nebraska.

Far from any paved roads west of Chugwater, Wyoming, in mid-June of 2016, mammatus clouds are illuminated by the setting sun with faint anticrepuscular rays visible beneath them.

Lightning and the setting sun illuminate the underside of a shelf cloud over the Cheyenne River Reservation in South Dakota during mid-June of 2018. The churning and ragged appearance is a result of the turbulent interface between cold, rain-cooled air rushing out from the storm and the warmer, less dense air it lifts out of the way.

Opposite above: Rays of sunlight beam through a gap in clouds near the edge of a low precipitation supercell west of Chugwater, Wyoming, in mid-June of 2016. The tornado risk is very low with this type of storm; however, they are known for excellent displays of storm structure and large hail.

Opposite below: Crepuscular rays and the setting sun light up this low precipitation supercell spinning over the hills west of Chugwater, Wyoming, as it takes on a characteristic "barber pole" appearance in mid-June of 2016.

A large shelf cloud approaches near Faith, South Dakota, in early June of 2018. Dirt roads like this often provide great views for storm chasers, but they can quickly become difficult to drive on or impossible to traverse after heavy rain. Storm chasers venturing onto dirt roads must stay ahead of storms to reduce the risk of getting stuck.

A dazzling display of mammatus clouds overhead and distant cumulus towers beneath them near Elburn, Illinois, in late April of 2022. While powerful thunderstorms often produce mammatus clouds like these, impressive mammatus clouds can also appear in the absence of severe weather.

Rain curtains can be seen beneath this striated storm near Alton, Kansas, during mid-June of 2014.

A large shelf cloud indicates the location of the gust front associated with this supercell above the Lake Traverse Reservation of South Dakota during mid-June of 2012.

Spinning like a top and sounding like a raging waterfall, this tornadic supercell fortunately stayed over the desolate landscape of the Nebraska sandhills during late May of 2013. This appearance of stacked saucers only lasted for seconds before becoming a twisting, bubbling mass of cumulonimbus clouds.

Looking north across cornfields at a supercell near Louisville, Nebraska, during mid-June of 2018.

The large wall cloud of a supercell near Louisville, Nebraska, during mid-June of 2018 just prior to producing a tornado. A wall cloud is lower than the rest of the supercell's cloud base because rain-cooled air from the storm's downdraft is being lifted in this region in addition to the warm and humid inflow air, resulting in the rising air reaching saturation closer to the ground in this portion of the storm's updraft.

A supercell intensifies after dark near Red Cloud, Nebraska, in mid-June of 2014 as a strengthening low-level jet makes the environment more favorable for supercells and tornadoes. This storm would go on to produce a few nighttime tornadoes.

Mammatus from a low precipitation supercell west of Chugwater, Wyoming, during mid-June of 2016. Faint anticrepuscular rays are visible to the right of the mammatus clouds near the horizon.

Bulging mammatus on the underside of an anvil cloud near Alton, Kansas, in mid-June of 2014.

Strong winds accompany this shelf cloud near Dell Rapids, South Dakota, in early June of 2020. This storm would produce severe winds along a path greater than 100 miles in length.

An elevated thunderstorm with striations near Grandin, North Dakota, during early July of 2018. With a stable layer of air near the ground, the air in the updraft of elevated thunderstorms like this one originates from higher up in the atmosphere, above the stable air that is too difficult to lift.

A turbulent and highly sheared shelf cloud approaches with an ominous turquoise coloring near New England, North Dakota, in late May of 2018.

Multiple inflow bands feed into this striated high precipitation supercell near Halstad, Minnesota, during mid-July of 2017. An improvement in the appearance of inflow bands, such as the ones on the right side of this storm, are often a visual indication of a strengthening storm.

Hay bales, possibly the ones in the foreground, were ultimately tossed thousands of feet by a tornado from this large supercell near Borup, Minnesota, in mid-July of 2017.

A large, low shelf cloud approaches near Rock Rapids, Iowa, in early June of 2016.

Cold and gusty winds overtake the area while looking underneath the length of a shelf cloud near Rock Rapids, Iowa, in early June of 2016. Looking almost like fingers reaching down from the storm, rising filaments of cloud indicate where the surging outflow is lifting warm air in its path up into the shelf cloud.

Multiple inflow bands stream into this supercell with a wall cloud beneath it near Oxford, Iowa, in mid-July of 2020.

A lack of strong surface winds going into this supercell allow cold downdraft air to quickly undercut the wall cloud in this photo near Oxford, Iowa, in mid-July of 2020. This limited the hazards from this storm to large hail and strong straight-line winds.

The updraft base of a young but intensifying storm is worth stopping to look at near Ransom, Kansas, in mid-May of 2013.

On the right, rain from the forward flank downdraft can be seen falling and creating a "rain foot." Small fragments of cloud called "scud" are visible moving from near the forward flank downdraft into the wall cloud in the center of this photo from late April of 2020 near Mendota, Illinois.

Explosive updraft growth of a thunderstorm in northwest North Dakota during mid-June of 2018. This storm was targeted by hail suppression aircraft in the region.

An intense cascade of hail and torrential rain creates an ominous white mass beneath this supercell near Wilmot, South Dakota in mid-June of 2012.

This storm near Perley, Minnesota, in early July of 2018 takes on a greenish hue.

An ominous lowering stretches down towards the ground behind a bobbing pumpjack in a field near Ness City, Kansas, in mid-May of 2013.

Near Ulen, Minnesota, this high precipitation supercell continues to produce tornadoes shrouded in rain during mid-July of 2017. While the tornadoes themselves cannot be viewed, the supercell provides impressive structure when viewed from afar.

The precipitation core containing damaging hail and heavy rains rapidly approaches from this supercell near Alexander, Kansas, in mid-May of 2013.

Drought conditions in Texas contributed to a dust storm forming in the inflow into this immense supercell near Morton, Texas, in late May of 2022. Reddish-orange dust provides a unique visual of the airflow in the vicinity of this monstrous storm that is typically invisible to storm chasers.

Air streaming into this supercell begins accelerating, and the layer of intense inflow, shown by blowing dust, tightens closer to the ground. The birth of what will ultimately become a nearly mile-wide tornado is underway near Morton, Texas, in late May of 2022.

4

TORNADOES

Although tornadoes have been documented in all fifty states and many countries, the level and frequency of tornado risk varies greatly across the United States and around the world. "Tornado Alley" seeks to define the geographic region that is a hotspot for tornado activity; however, depending on which details you care about most (frequency, intensity, and impacts, just to name a few), the boundaries you select for "Tornado Alley" may differ significantly. For storm chasers, this colloquialism often highlights the Great Plains and portions of the Midwestern United States, where low population density, excellent visibility, and some of the highest tornado counts on the planet provide the best chance at witnessing the potentially photogenic moments when a rapidly rotating column of air connects the gap between the cumuliform clouds of the storms above and the ground below.

A large tornado near Morton, Texas, in late May of 2022 emerges from dust being carried by the inflow of the storm. Persistent drought conditions caused typically favorable viewing locations to be swallowed up in thick dust for this storm.

As a gap in blowing dust develops, this wedge tornado, nearly one mile wide, can be seen near Morton, Texas, in late May of 2022. Fortunately for nearby towns, this strong tornado remained over open, dusty fields during its entire life.

Hail and rain fall in the foreground of a multi-vortex tornado northwest of Sterling City, Texas, during mid-May of 2021. At its peak width, this tornado exceeded a half-mile in diameter.

Suction vortices of this multi-vortex tornado during mid-May of 2021 tower above wind turbines rising from the Texas countryside. The areas impacted by these suction vortices, also known as secondary vortices, often suffer the worst damage from a tornado.

As swirling rain curtains begin surrounding this mid-May 2021 tornado in Texas, one of the suction vortices connects the large cone funnel above to the ground. These fully condensed suction vortices often appear and vanish within seconds while rotating around the tornadic circulation.

While suction vortices dancing beneath the large cone funnel of this tornado indicate the danger present, not all tornadoes will have a fully condensed funnel or visible suction vortices stretching to the ground. Just like with this mid-May 2021 Texas tornado, tornadic winds often exist in cloud-free air underneath the funnel above.

A tornado develops from a wall cloud near Morris, Minnesota, during mid-June of 2017. Fortunately, this tornado largely remained over open fields, and only produced limited damage just prior to dissipating.

Different vantage points of tornadoes often bring very different lighting. Looking east at this mid-June of 2017 tornado near Morris, Minnesota, it appears dark gray; however, those looking west at this same tornado saw it as bright white.

A distant wedge tornado north of Benkelman, Nebraska, during late May of 2021. The unusual north-northwest motion of the tornado along a boundary and sparse, often muddy roads made getting a closer look difficult.

With the shelf cloud from the rear flank gust front above, sunlight illuminates a tornado northwest of Sterling City, Texas, bright white during mid-May of 2021. Shortly after this photograph, the tornado disappeared behind rotating curtains of rain.

Dust and debris swirl at the base of this tornado near Morris, Minnesota, during mid-June of 2017. Just a few moments later this tornado would dissipate after taking a path primarily over open fields.

A distant stovepipe tornado rolls across open fields near Binford, North Dakota, during late June of 2015.

The silhouette of a large tornado wrapped in rain northeast of Mayville, North Dakota, during mid-July of 2017 before it quickly vanished behind curtains of heavy rain. Tornadoes associated with high precipitation supercells like this one are the most dangerous for storm chasers and the general public alike, as tornadoes can be lurking unseen behind a wall of heavy rain and hail.

Moments later, any sight of the ongoing, large tornado near Mayville, North Dakota, during mid-July of 2017 is gone. This large, rain-wrapped tornado would track 23 miles in total. Despite lasting almost an hour, for much of its life the tornado was completely hidden within rotating curtains of heavy rain and hail.

A cone tornado harmlessly moves across fields near Granville, North Dakota, during late July of 2013.

A view of a tornado during late July of 2013 near Granville, North Dakota, after a short jog by foot down the minimum maintenance road in the foreground. The lack of passable roads prevented getting a closer look at this tornado, but the ground circulation was confirmed by pilot reports.

On the left, dust is lofted by the rear flank downdraft of this supercell. Sunlight pours through the "clear slot," an area where clouds are eroded by the rear flank downdraft, to the left of this developing tornado near McCook, Nebraska, in mid-May of 2019.

A condensation funnel begins growing towards the ground through a rotating column of dust as this mid-May 2019 tornado near McCook, Nebraska, becomes increasingly photogenic.

Dust kicked up by the rear flank downdraft stretches across the horizon to this tornado near McCook, Nebraska, during mid-May of 2019. A prominent clear slot is visible to the left of the tornado where sunlight passes through the cut of cloud-free air in the storm.

Dramatically roping out, the condensation funnel narrows and twists as it is swallowed up in dust lofted by the powerful rear flank downdraft of this mid-May of 2019 supercell near McCook, Nebraska.

A wedge tornado churns across the open countryside near Benkelman, Nebraska, in late May of 2021 while a pump jack nods in the foreground. At times, the inside of this tornado appeared hollowed out and transparent like in this photo.

Opposite: This farmstead is spared the worst as this mid-May of 2019 tornado near McCook, Nebraska, narrowly misses it to the west. Dust swirls up the condensation funnel, illuminated by light pouring through the clear slot.

This large wedge tornado visually reaches its maximum width—fortunately over the sparse and open terrain—near Benkelman, Nebraska, during late May of 2021. An unusual north-northwesterly motion was observed with this tornado as it appeared to ride along a pre-existing boundary apparent on radar.

Near Max, Nebraska, in late May of 2021, cows turn their backs on this tornado churning through the hilly terrain.

Irrigation equipment and bales of hay near Max, Nebraska, fill the foreground as a strong tornado lurks in the distance during late May of 2021.

A powerful suction vortex can be seen at the bottom of this tornado near Benkelman, Nebraska, during late May of 2021. Closer, on the left, another rotating wall cloud has developed and will go on to produce a tornado as well.

Looking across Swanson Lake, in southwest Nebraska, two distant tornadoes are ongoing during late May of 2021. A third tornado from this same supercell, not seen in the photograph, was also simultaneously ongoing at this time.

This tall, white, rope tornado is remarkably persistent northwest of Max, Nebraska, during late May of 2021.

Fully condensed from the ground to cloud bases high above, this unique rope tornado northwest of Max, Nebraska, during late May of 2021 provides exceptional variety to a day that began with large wedge tornadoes.

In a seemingly never-ending rope out stage, a distant tornado on the right, north of Benkelman, Nebraska, continues while the next tornado is about to develop from the wall cloud on the left during late May of 2021.

A bizarre, cylinder-shaped tornado west of Max, Nebraska, during late May of 2021. While each tornado tends to look unique in its own way, the strange shape of this tornado is distinctly different from any others I have seen to date.

Despite an ominous and large rotating wall cloud, this tornado east of Okmulgee, Oklahoma, was fortunately brief and short-lived during late May of 2019. Not every tornado will have a well-defined condensation funnel stretching to the ground, and this one was only discernible from close range by the swirling dirt and debris under the rapidly rotating wall cloud.

A rope tornado twists through the sky near Paducah, Texas, during late May of 2019. Fortunately, this tornado missed the town, and the day overall did not live up to the fears and expectations of potentially being a historic tornado outbreak for Texas and Oklahoma.

During mid-May of 2019, a brief tornado kicks up dust near Stockville, Nebraska.

An impressive and imposing wall cloud produces a small tornado near Louisville, Nebraska, during mid-June of 2018. Straight-line winds with this storm ultimately ended up being the bigger story, causing far more significant damage than the tornado in this photo.

Across a soybean field, a tornado near Sycamore, Illinois, strikes a farmhouse and begins to loft debris during early August of 2021.

Pieces of debris are centrifuged outwards from this tornado near Sycamore, Illinois, during early August of 2021. In close proximity to tornadoes, falling debris is a significant danger, even in areas that may ultimately end up out of the path of the tornado itself.

Another shot of this photogenic tornado near Sycamore, Illinois, during early August of 2021. Small debris is lofted both high and far away from the tornado itself.

The damaging vortex at the surface appears to disconnect from the funnel above while debris scatters across the sky during early August of 2021 near Sycamore, Illinois. It is always important to maintain situational awareness and continually keep eyes on all parts of a storm, as the behavior of tornadoes can often be unpredictable and surprising like the scene in this photograph.

A cone tornado rapidly appears within the rotating rain curtains of the bear's cage northwest of Sterling City, Texas, during mid-May of 2021. Within the bear's cage—the region beneath the mesocyclone—tornadoes like this one can develop with little to no warning. Venturing into this part of the storm is a risky maneuver even for experienced storm chasers, especially if large rain curtains are present, which obstruct visibility.

This large, dusty tornado was rated an EF-3 during mid-May of 2019 northeast of Stockville, Nebraska. Tornadoes are rated from EF-0 (weakest) to EF-5 (strongest) based on damage surveys. This rating system can cause bias for sparsely populated areas of the country where tornadoes typically cause little or no damage, primarily because they remain over open fields. EF-U ratings are often assigned to these tornadoes now with "U" standing for an "unknown" intensity due to the lack of damage indicators in the path of the tornado.

A brief, weak tornado forms on the southwest side of a storm near Canova, South Dakota, in mid-May of 2015. Despite being chilly for a storm chase, abundant low-level instability and surface vorticity were enough for a tornado to form with this storm.

240 ST

East of Victoria, Illinois, during late June of 2022, this short-lived tornado within a line of storms provides a quick photo opportunity while tracking over an unpopulated area.

Opposite above: The unusual location of this tornado within a storm near Canova, South Dakota, during mid-May of 2015 allows it to be brightly lit up by the sun and appear bright white. This tornado quickly roped out after coming within a few hundred yards of a farmhouse.

Opposite below: A weak tornado becoming rain wrapped near Kirkland, Illinois, during early August of 2021. This was one of many tornadoes that afternoon in Northern Illinois, on a day where several storms each produced multiple tornadoes.

5

LIGHTNING

Wide-open terrain, limited light pollution, and frequent thunderstorm activity make Tornado Alley one of the most exciting places to see lightning. With that excitement comes danger and fear for many storm chasers. Unlike a tornado or large supercell that you can see approaching or track with radar data, lightning happens in an instant and often without warning, posing the greatest danger to most storm chasers aside from the inherent risks of driving thousands of miles each spring in pursuit of storms. Complex processes within thunderstorms lead to charge separation, and the resulting lightning can be thought of as similar to a massive version of the sparks of static electricity that you have likely felt at some point, perhaps touching a doorknob after walking across a carpet. The static shock of lightning, however, is capable of traveling tens of miles through the atmosphere with enough power to explode the bark off of a tree and set it on fire in a split second.

Multiple bolts of cloud-to-ground lightning strike in the distance beyond a soybean field near Elburn, Illinois, in early July of 2022.

A vivid bolt of lightning strikes near Elburn, Illinois, illuminating sheets of rain in early July of 2022.

A forked bolt of lightning strikes near Friend, Nebraska, in early June of 2020. The brightest of the branches is the one that connects the cloud and the ground.

Lightning strikes west of Grand Forks, North Dakota, in early July of 2018. Multiple faint stepped leaders that did not connect with the ground are visible in addition to the bright path of the return stroke where a stepped leader reached the ground.

The isolated nature of this storm near Fort Morgan, Colorado, allows both stars in the upper right and bolts of lightning to be captured in a single photo during late June of 2016.

Curtains of falling rain are illuminated with these lightning strikes, again captured along with stars of the night sky near Fort Morgan, Colorado, during late June of 2016.

Complex bolts of lightning partially masked by low clouds strike beyond a railroad crossing near Friend, Nebraska, in early June of 2020.

An intense bolt of lightning through turbulent skies beyond a railroad crossing makes for a dramatic scene near Friend, Nebraska, in early June of 2020.

Look closely at the power lines. This complex forked bolt of lightning has branches both in front of and behind the power lines near Emerado, North Dakota, in late August of 2012.

A large bolt of lightning strikes near Lincoln Drive Park in Grand Forks, North Dakota, during early June of 2017.

Vivid colors from the sun setting beneath this supercell near Grand Island, Nebraska, surround frequent lightning during early June of 2020.

A brief funnel cloud is lit up by a cloud-to-ground lightning strike near Colby, Kansas, in mid-June of 2012.

Lightning spreads across the sky near Anselmo, Nebraska, in late May of 2013.

Spanning tens of thousands of feet, this lightning bolt strikes near Anselmo, Nebraska, in late May of 2013. You would not have to ask a storm chaser twice to stop and take a look at this storm's display of lightning.

From this rapidly growing updraft, a bolt of lightning strikes the open terrain near Raymer, Colorado, in late May of 2020. This was just the beginning of a nearly constant barrage of lightning lasting for hours from this storm.

On July 4, 2022, nature puts on a firework show with hours of lightning to go along with occasional informal fireworks displays like those seen on the lower right near Lily Lake, Illinois.

6

SKY AND ATMOSPHERE

The absence of severe weather does not limit the awesome appearance of the skies above Tornado Alley. Beautiful terrain and endless views of the sky allow planning and positioning to view rainbows, dramatic sunsets, and the night sky away from the glare of city lights. The northern fringes of Tornado Alley also offer an occasional chance at viewing the aurora borealis when clear skies are present. Lastly, brief moments of chance and turbulence in the atmosphere can lead to unique cloud formations and spectacular lighting for storm chasers even on days that may have begun with low expectations.

If the sun is low enough in the sky, viewing a storm with the sun at your back offers a chance at seeing rainbows during the late afternoon like this one near Warren, Minnesota, in early July of 2018. The lower the sun is in the sky, the higher the rainbow will be above the horizon.

A faint double rainbow appears as a storm dissipates over a wind farm near Mayville, North Dakota, during late June of 2015.

The setting sun casts a variety of colors on mammatus overhead from a supercell near Chugwater, Wyoming, during mid-June of 2016.

This bright rainbow lit up turbulent skies around sunset near Hillsdale, Illinois, during mid-July of 2020. On days when the severe weather threat diminishes towards evening, positioning to see rainbows on the backside of a storm can be an exciting way to end a storm chase.

Horseshoe vortex clouds like this one require a precise balance of an updraft interacting with wind shear, and usually only last for a few seconds if they form. This example found in the inflow of a prolific tornado producing supercell near Morton, Texas, was remarkably persistent and lasted for minutes during this late May evening of 2022.

A rainbow develops from misty skies near Medora, North Dakota, in early September of 2013.

Unexpected chaotic events often create memorable scenery like this beam of light shining through gaps in clouds and precipitation from distant storms near Colby, Kansas, in mid-June of 2012.

Positioning for a rainbow pays off. A long lasting and intense double rainbow develops near Norris, South Dakota, in late May of 2022.

Turbulent skies colored by the setting sun over the rugged terrain near Medora, North Dakota, in early September of 2013.

In the open prairies of Tornado Alley, sometimes fair-weather days turn into photoshoots like this late June 2012 sunset with spectacular colors near Amidon, North Dakota.

When chasing the northern portions of Tornado Alley, it is important to also keep track of any geomagnetic storms that might be underway. A geomagnetic storm in late June of 2015 produced the aurora borealis display seen here, one that was so bright it was visible before the sunset colors had faded near Manvel, North Dakota.

The familiar yellow-green color of the aurora borealis comes from the excitement of oxygen about 60 miles above the Earth's surface. The excitement of oxygen at higher altitudes can lead to red auroras, and the presence of nitrogen in the atmosphere can even result in shades of blue or shades of purple under the right conditions. This late June 2015 aurora near Manvel, North Dakota, shows a brilliant display of greens, reds, and even purple.

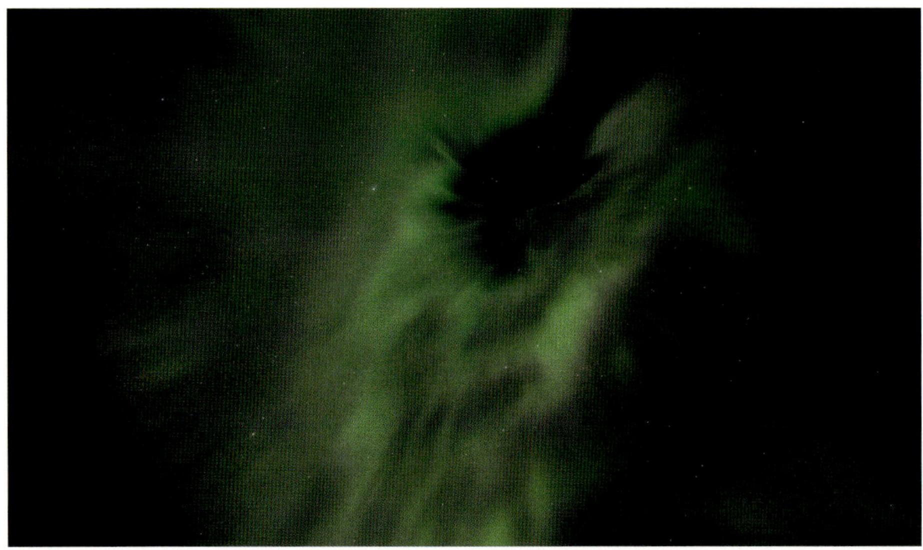

Looking directly overhead at a pulsing auroral display near Manvel, North Dakota, during late June of 2015. This beautiful summertime display was a welcome change of pace from the more typical subzero temperatures that often accompany viewing the aurora borealis in this part of the world.

A rare instance of storm chasers trying to find as few clouds as possible to view the total solar eclipse near Alliance, Nebraska, in late August of 2017. While storm chasers regularly bring significant economic activity to the towns and states of Tornado Alley, the solar eclipse was estimated to have brought over 600,000 out-of-state visitors to Nebraska along with over 100 million dollars of economic activity.

Tranquil clear nights provide excellent astrophotography opportunities throughout the dark skies of Tornado Alley. On this early June of 2021 night in Henderson, Nebraska, the milky way is visible above an alfalfa field next to irrigation equipment.

Just south of Henderson, Nebraska, in early June of 2021, the milky way stretches across the night sky above grain bins. Sights like this are easily attainable in Tornado Alley for those willing to stay out late away from city and town lights.

7

STORM CHASERS

Each severe weather season, thousands of people from all over the world travel countless miles for a chance to witness firsthand the stunning skies, landscapes, and storms of Tornado Alley. Even with quiet weather patterns, memories of a lifetime are made exploring the small towns, unique landmarks, and local businesses that speckle the vast, endless landscapes of the open plains. Storm chasers are an eclectic mix, ranging from self-taught hobbyists, professional videographers, and tour groups to research scientists, college field studies classes, and even professional meteorologists on "chasecations." Storm reports that chasers provide assist the National Weather Service in issuing accurate and timely warnings, and the observations of chasers also help in assessing the accuracy of the forecasts and warnings that were issued. While the motivations and goals for individual storm chasers can often vary greatly, the common passion for experiencing and trying to better understand the wild weather of Tornado Alley creates a strong sense of community among this very diverse group of people.

A research vehicle from the National Severe Storms Laboratory approaches a tornado near Farnham, Nebraska, in mid-May of 2019 collecting data as part of the TORUS research project. Targeted observations are still critical in improving our understanding of supercells and the tornadoes they can produce.

Battling dusty inflow, the College of DuPage field studies class views the structure of a supercell near Morton, Texas, in late May of 2022. This view came shortly after witnessing a wedge tornado through a break in the inflow dust.

After years of monsoon chasing and lightning photography in the southwest, Flano tries out chasing in Tornado Alley and sees his first tornado near McCook, Nebraska, in mid-May of 2019. Since then, Flano has witnessed over a dozen tornadoes and works in Tornado Alley chases whenever he can.

On his first day chasing with the College of DuPage, Brad gets his first tornado near McCook, Nebraska, in mid-May of 2019. In the time since this first tornado, Brad has witnessed over a dozen additional tornadoes.

Dr. Gilmore underneath mammatus clouds during mid-May of 2017 while leading the University of North Dakota storm experience class. Throughout his career, Dr. Gilmore has done extensive research on tornadoes and microphysics, and continues to share his passion for severe weather research with students in classes like this one.

Margo, a meteorology graduate student and trip assistant for College of DuPage storm chases, hikes to a high point in Theodore Roosevelt National Park in mid-June of 2022. Fair weather days like this one provide storm chasers with an opportunity to explore the sights and activities of the vast territory they traverse each year in pursuit of storms.

Opposite above: Ana, an EMT, and Jim, a disaster resource coordinator and former EMT, are always prepared to serve as first responders if needed when out chasing. Jim has become a regular driver for College of DuPage storm chasing trips, as well as an adjunct faculty member for weather hazards and preparedness courses.

Opposite below: Evan (left) and Mike (right) keep NEXLAB at the College of DuPage running strong, while continually developing and improving weather.cod.edu to best support the needs of the meteorology and storm chasing programs at College of DuPage. Thousands of meteorologists and storm chasers around the world also turn to weather.cod.edu for critical weather information, with the satellite and forecast models pages serving as established favorites for many during the storm chasing season.

F150

True-Color Imagery for Full Disk (Large) (GOES-East)
College of DuPage
NEXLAB Satellite and Radar
Select a Sector Category:
View Global Sectors
Select a Product:
BenQ

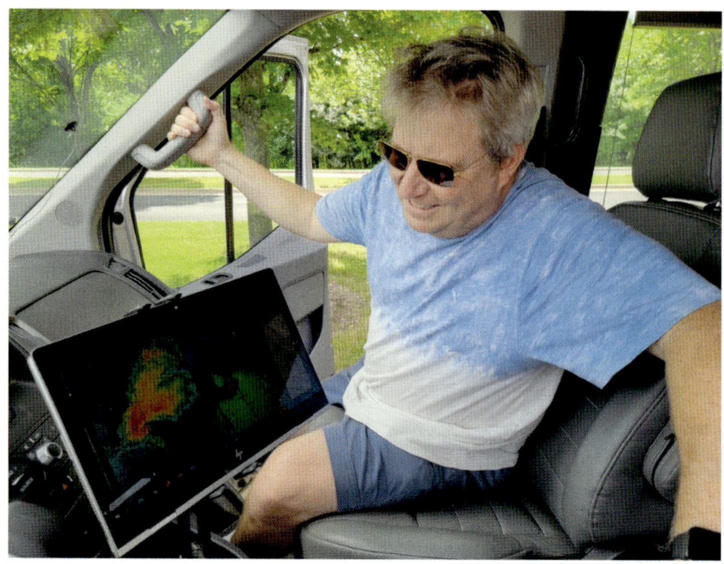

Starting the first college storm chasing program over thirty years ago, Paul has created an unforgettable educational experience for students that numerous universities have modeled their own storm chasing programs after. Additionally, Paul has been the driving force behind creating and developing weather.cod.edu to support the meteorology curriculum at the College of DuPage, a website which has become a critical piece of the meteorology community at large.

From brush guards, KC lights, and radio setups, to the numerous demands of turning vans into mobile storm chasing classrooms capable of weather balloon launches, Leon finds a way to design, engineer, and build whatever is needed to support the College of DuPage storm chasing program. A veteran driver for College of DuPage storm chasing trips, Leon has helped take hundreds of students on the adventure of a lifetime.

Myself, with a large tornado near Morton, Texas, in late May of 2022. After spending the afternoon in Morton under blue skies waiting for storms, our group saw the development of this impressive supercell followed by the strong tornado that harmlessly spun over dusty fields. The fortunate lack of damage and sharing this experience with my class make it one of my all-time favorite chases.

Aaron, a PhD candidate at the University of North Dakota at the time of this photo, was working as a trip assistant helping get the College of DuPage group in position to witness this wedge tornado near Morton, Texas, in late May of 2022.

Deanna (left) is a criminal justice professor at the College of DuPage and a longtime COD storm chasing driver who has driven countless miles bringing classes into position to view dozens of tornadoes and hundreds of supercells. Christene (right) is thrilled to see her first tornado since her teenage years, especially after the hard work she has put in the past few years taking college meteorology courses to inform her passion for severe weather and storm chasing.

Working in finance for the city of Champaign, Joe has taken time each storm season to drive for the College of DuPage storm chasing program. Frequently the "van 1" driver, Joe has been leading the way on the roads to and from many tornadoes and photogenic storms for over fifteen years.